THE WATER CYCLE™

WATER IN PLANTS AND ANIMALS

Isaac Nadeau

The Rosen Publishing Group's
PowerKids Press™
New York

To Evergreen

Published in 2003 by The Rosen Publishing Group, Inc.
29 East 21st Street, New York, NY 10010

Copyright © 2003 by The Rosen Publishing Group, Inc.

All rights reserved. No part of this book may be reproduced in any form without permission in writing from the publisher, except by a reviewer.

First Edition

Editor: Gillian Houghton
Book Design: Maria E. Melendez

Photo Credits: Cover, title page, and all page borders © EyeWire; cover and title page, deer © Digital Stock; pp. 4, 7, 8, 11, 12, 15, 16, 19 (right), p. 20 Maria E. Melendez; p. 19 (left) © LifeArt/TechPool Studios, Inc.

Nadeau, Isaac.
 Water in plants and animals / Isaac Nadeau.
   p. cm. — (The Water cycle)
 Includes bibliographical references (p.   ).
 ISBN 0-8239-6264-4 (lib. bdg.)
 1. Water—Juvenile literature. 2. Plant-water relationships—Juvenile literature. 3. Animal-water relationships—Juvenile literature. 4. Hydrologic cycle—Juvenile literature. [1. Water. 2. Hydrologic cycle.] I. Title.
 GB662.3 .N34 2003
 572'.5394—dc21

2001006170

Manufactured in the United States of America

CONTENTS

1	Water Everywhere	5
2	Water and Life	6
3	From Soil to Roots	9
4	Turning Water into Food	10
5	Back into the Air	13
6	Plants in Dry Places	14
7	Wetlands	17
8	How Water Moves in Animals	18
9	Animal Adaptations	21
10	The Water Cycle in Us and Around Us	22
	Glossary	23
	Index	24
	Web Sites	24

atmosphere

ocean

river

lake

WATER EVERYWHERE

All of the water on Earth is part of the water cycle. The water cycle is the **circulation** of water throughout the planet. A drop of water can be frozen in ice for a thousand years before melting and flowing downhill to the ocean. Ocean water **evaporates** and is carried by air currents into the atmosphere, where it might become part of a cloud. Before long the drops of water fall back to Earth as rain or snow. They might sink into the ground, where the roots of plants **absorb** water from the soil. They might collect in a pond and be lapped up by a thirsty raccoon. The water in the roots and the stems of plants and the water in a raccoon's body is an important part of the water cycle.

There is about the same amount of water on Earth today as there was a million years ago. This is because water is constantly being recycled through the water cycle. Water can be polluted, but scientists believe that it can never be used up.

WATER AND LIFE

Just as water moves throughout Earth, it also circulates in the bodies of living things. All living things are made up of tiny units called cells. The smallest living things are made up of only one cell. Larger plants and animals can have billions of cells. Cells give living things their structure. They also help to carry out all of the functions a living thing needs to stay alive, including breathing and carrying **nutrients** throughout the body. Cells depend on water for these important jobs. All living things are made up of **molecules**. Most of these molecules are water. A molecule of water is made up of two gases, **hydrogen** and oxygen. Hydrogen and oxygen are two of the most important nutrients for living things.

The human body is about 65 percent water. A tomato plant is about 80 percent water, and the tomato itself is 95 percent water. Water is made up of two parts hydrogen and one part oxygen (inset).

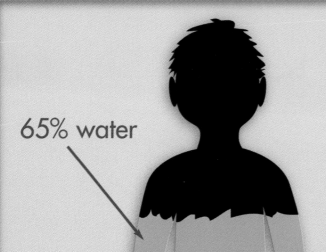
65% water

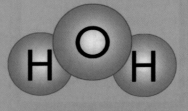
the water molecule

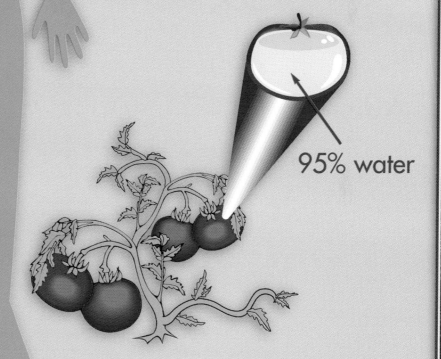
95% water

FROM SOIL TO ROOTS

When rain falls to the ground, some of the water soaks into the soil. This process is called **infiltration**. The underground water absorbs nutrients from the soil. Plants use their roots to draw this nutrient-rich water out of the soil and up into their stems and leaves. Water moves from cell to cell inside the bodies of plants, bringing nutrients with it. There often is a limited supply of water and nutrients in the soil, so different kinds of plants have roots that grow to different depths under ground. Some plants have shallow roots that collect water near the surface. Some plants have long, thick roots, called taproots, to gather water from deep under ground.

◄ Plant roots grow from the tips, stretching toward the best places to find water and nutrients in the soil. When the ground is saturated, or filled, with water (inset), the roots of a plant absorb the water. Arrows show the path of water from the roots to the leaves.

TURNING WATER INTO FOOD

Water plays a key role in **photosynthesis**. Photosynthesis is the process in which plants make food out of sunlight, water, and air. The cells of leaves have water in them. Many of these cells also have **chloroplasts**, specialized parts of cells where photosynthesis takes place. Sunlight shines on the leaves of plants all day. Chloroplasts use the heat from sunlight to split the water molecules apart, separating the hydrogen from the oxygen. The oxygen escapes the surface of a leaf as a gas that people and other animals need to breathe. At the same time, the leaves absorb **carbon dioxide** from the air. In the chloroplasts, hydrogen is combined with carbon dioxide to make sugar. This sugar is the plant's food. Water carries it throughout all the cells of the plant, even the tips of the roots.

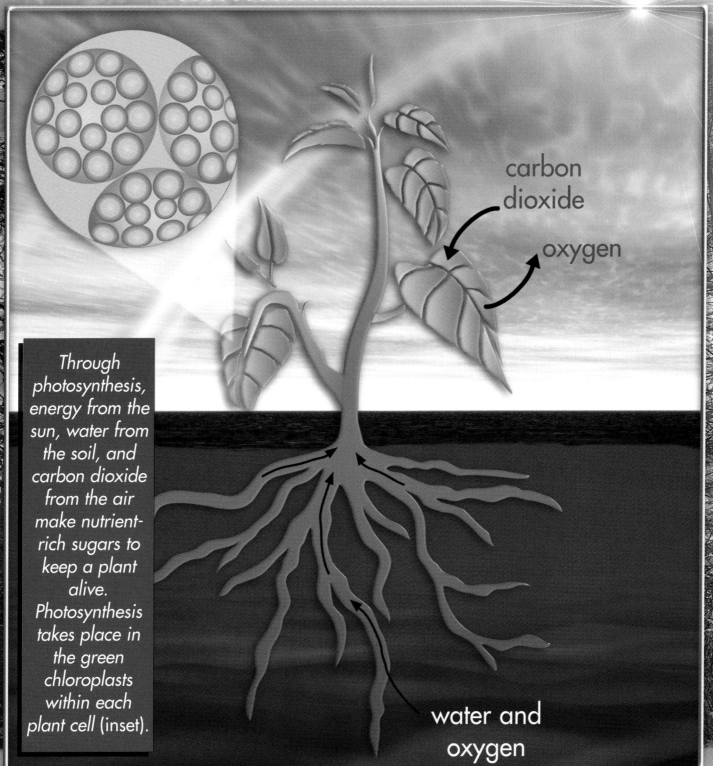

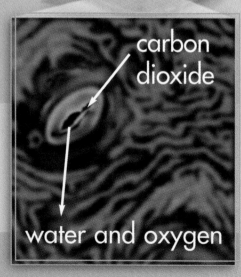

carbon dioxide

water and oxygen

If there is not enough water in the soil to replace the water lost through transpiration, a plant will close its stomata (inset) and begin to wilt. This slows down the rate of transpiration, but it also slows down the process of photosynthesis.

BACK INTO THE AIR

Plants lose a lot of water through their leaves. Every leaf is dotted with thousands of tiny surface openings called stomata. The word "stomata" comes from the Greek word for "mouths." These mouths are where gases, such as oxygen, carbon dioxide, and water vapor, are exchanged with the surrounding air. The stomata absorb carbon dioxide from the air and allow oxygen to escape. Most of a plant's water escapes from the stomata, too. The evaporation of water from the leaves of a plant is called **transpiration**. Water molecules stick together, creating a chain of water from the roots to the leaves. As each molecule evaporates from a leaf, another molecule is pulled up the chain to replace it. In tall trees, water can travel more than 300 feet (91 m) from the tips of the tree's roots to its highest leaves.

PLANTS IN DRY PLACES

Plants can grow almost anywhere on land. They can grow in the driest and the wettest conditions because of special **adaptations**. Adaptations are changes that have occurred in a living thing that make it suited to its environment. Many deserts receive fewer than 12 inches (30 cm) of rainfall per year. Plants that live in dry places have adaptations that allow them to conserve, or save, the little water that is available to them. Some plants, called **succulents**, gather a lot of water when it rains and store it in their roots, stems, or leaves for use during dry seasons. Some plants open their stomata only at night, when transpiration occurs much more slowly. At night these plants take in carbon dioxide from the surrounding air through their stomata. They save it until daytime, when it can be used in photosynthesis.

The accordion-like ribs of the saguaro cactus (above) allow the plant to swell when there is a lot of water available and to shrink when it is dry. The agave (inset) is a succulent, not a cactus. It has narrow, fleshy leaves that hold in water.

The roots of a mangrove tree grow above the surface of the water to get oxygen from the air. In some wetlands, such as bogs, the soil might not contain all of the nutrients a plant needs. Venus's-flytraps (inset) get important nutrients by trapping and digesting insects in their mouthlike leaves.

WETLANDS

A wetland is a place that has water on or just below the surface of the land for at least part of the growing season of the plants that live there. Ponds, marshes, bogs, and swamps are all examples of wetlands. They differ in how much water they hold, what kind of soil they have, and what kinds of plants live in them. The plants, called **hydrophytes**, that live in wetlands have adapted to wet conditions. Plants give off oxygen to the air during photosynthesis, but they also need to take in oxygen through their roots to help them grow. In wetlands the soil can be so filled with water that little air or oxygen can reach the roots. Some plants, such as mangroves, have unusual roots that grow above the surface of the water. These roots act like **snorkels**, taking oxygen out of the air for the roots in the water below.

HOW WATER MOVES IN ANIMALS

Water is important to animals, too. Every living cell in an animal's body has water in it. When you work or play, your body sweats. This sweat, or perspiration, evaporates from your skin and becomes water vapor in the air. Breathing, or respiration, also adds water vapor to the air. Perspiration and respiration help many animals to keep cool. Water also leaves the body through **urine**, carrying away waste. Like plants, animals need to replace the water they lose.

Inside the bodies of humans and other animals, water is constantly moving from one place to another. Water is the main ingredient in blood, which carries oxygen and other nutrients throughout an animal's body. **Blood vessels** connect all of the parts of the body to the heart. The heart pumps blood through the blood vessels.

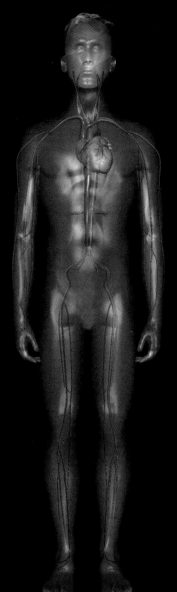

A human being drinks about 16,000 gallons (60,000 l) of water in a lifetime! Inside the body, blood carries water to the organs. The heart pumps blood through a complex system of blood vessels (left). Arteries (in red) carry blood away from the heart, while veins (in blue) carry blood to the heart.

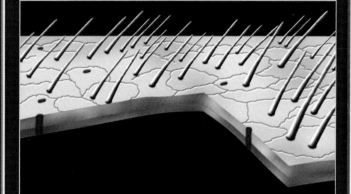

There are about 10 hairs and 100 sweat glands in every square centimeter (.16 sq in) of human skin (above). Our skin controls the balance of water in our bodies. Skin holds most water in; but it allows some water, in the form of sweat, to leave the body.

The desert-dwelling kangaroo rat gets all of the water it needs to survive from the cells in the seeds it eats. It never has to take a drink!

ANIMAL ADAPTATIONS

Animals live in all kinds of conditions. They live in places that are very dry, such as deserts, and places like wetlands that are very wet. No matter where they live, all animals need water. Many desert animals have adapted to their hot surroundings by being active only at night. These animals lose less water from perspiration, because their bodies sweat less in the cool, night air. They also lose less water through respiration, because the moist night air replaces much of the water vapor that is lost as the animals breathe.

Many animals are unable to drink ocean water, because it is too salty. Salt draws water out of cells in an animal's body, preventing the water from being absorbed. Albatrosses, giant birds that fly for miles (km) over the open oceans, have a special **gland** that removes the extra salt from seawater.

The Water Cycle In Us and Around Us

The water cycle is necessary to all life on Earth. In turn life itself is part of the water cycle. All living things are connected to oceans, rivers, rain, and one another by the water cycle. The water you drink today might have been in the roots of a wetland plant a year ago. The same water might have been in the cells of a dinosaur, and someday your grandchildren might drink it. It is important for us to care for our water so that it is clean for all the living things that depend on it. In some cases, wetland plants help to clean polluted water. Duckweed, for example, is able to absorb many pollutants from the water. Many cities have learned the role wetland plants play in keeping the water clean for all the plants and the animals that use it.

GLOSSARY

absorb (uhb-ZORB) To take in and hold on to something.
adaptations (a-dap-TAY-shuns) Changes made to fit conditions.
blood vessels (BLUD VEH-suhlz) The tubes in the body that carry blood.
carbon dioxide (KAR-bin dy-OK-syd) A gas that the body makes to get rid of waste from energy that was used.
chloroplasts (KLOR-uh-plasts) The parts in plant cells where photosynthesis takes place.
circulation (sir-kyoo-LAY-shun) The circular movement of something.
evaporates (ih-VAH-puh-rayts) Changes from a liquid to a gas.
gland (GLAND) An organ or a part of the body that produces a substance to help with a bodily function.
hydrogen (HY-druh-jihn) A gas that is common in the atmosphere.
hydrophytes (HY-druh-fyts) Plants that live in very wet conditions.
infiltration (ihn-fihl-TRAY-shun) Water soaking into the ground.
molecules (MAH-lih-kyoolz) The tiny building blocks that make up a substance.
nutrients (NOO-tree-ints) Anything that a living thing needs for its body to live and grow.
photosynthesis (foh-toh-SIN-thuh-sis) The process in which plants use energy from sunlight, gases from air, and water to make food and to release oxygen.
snorkels (SNOR-kuhlz) Plastic tubes that run from the mouth to above the surface of the water, allowing a person to breathe under water.
succulents (SUHK-yoo-lehnts) Plants that store water in their roots, stems, or leaves for use during dry periods.
transpiration (tranz-puhr-AY-shun) Evaporation from the leaves and the stems of plants.
urine (YER-in) A form of liquid waste from the bodies of animals.

INDEX

A
adaptations, 14

C
carbon dioxide, 10, 13–14
cell(s), 6, 10, 18
chloroplasts, 10

D
deserts, 14, 21

H
hydrogen, 6, 10
hydrophytes, 17

I
infiltration, 9

L
leaves, 10, 13

N
nutrient(s), 6, 9

O
oxygen, 6, 10, 13, 17

P
perspiration, 18, 21
photosynthesis, 10, 14, 17

R
rain, 5, 9, 14
respiration, 18, 21
roots, 9–10, 13, 17, 22

S
stomata, 13–14
succulents, 14

T
transpiration, 13–14

W
water vapor, 18, 21
wetland(s), 17, 21–22

WEB SITES

To learn more about water in plants and animals, check out these Web sites:

www.epa.gov/rgytgrnj/kids/wets_b.htm
www.wetland.org/kids/Kids.htm